THE COAL BUSINESS

ON THE

PENNSYLVANIA RAILROAD.

A COMMUNICATION ADDRESSED TO

THE PRESIDENT, DIRECTORS, AND STOCKHOLDERS OF

THE PENNSYLVANIA RAILROAD, ON THE

COST OF TRANSPORTATION.

PHILADELPHIA:
T. K. AND P. G. COLLINS, PRINTERS.
1857.

INTRODUCTORY.

PHILADELPHIA, February 3, 1857.

*To the President, Directors, and Stockholders
of the Pennsylvania Railroad Company.*

HAVING been instrumental in the organization of various coal and lumber companies now in operation, or about to commence operation on or near the Pennsylvania Railroad; having exerted myself for years to encourage the investment of capital in such companies for the purpose of establishing a business which I considered essential to the success and satisfactory operation of your road, I hope that no apology will be required for this communication. A period of service embracing nearly ten years in the various offices of Assistant Engineer, Superintendent, Chief Engineer and Director, has now terminated, and my connection with the road has ceased. I have also withdrawn either entirely or partially from several of the companies which I assisted in organizing. *Personal interests* have not suggested this communication; but I still retain a desire to see the Pennsylvania Railroad become an instrument of more extended utility to the public, a means of further developing the resources of the State, and a source of ample remuneration to its stockholders. I feel also under imperative obligations as far as possible to protect those who have invested capital on my recommendation from the consequences of the erroneous views which are prevalent in reference to the cost of transportation upon the Pennsylvania Railroad.

This communication has been addressed to the Stockholders as well as to the President and Directors, because the opinion is prevalent amongst them that no profit can be derived from the transportation of coal on the Pennsylvania Railroad; in fact, it has been repeatedly asserted that coal is carried at a loss to the Company, to advance the interests of individuals; and so often has this declaration been made, so widely has it been disseminated by newspaper publications and otherwise, that there is reason to believe that it has been, and still is, credited by many whose good opinions are of value, and who, by entertaining such views, do much injustice to a Board of Directors who have uniformly gone quite to the opposite extreme, and have curtailed the business and profits of the Company by pertinaciously adhering to higher charges for transportation than the market value of the article will bear.

In the following pages I do not expect any consideration in consequence of past services or experience. I desire that no assertion shall be credited unless it be sustained by proof, that no results be admitted unless the process of reasoning or of calculation by which they are obtained is so simple and intelligible that you can be satisfied both with the premises and conclusions. I will endeavor to make perfectly plain and obvious a subject which must be admitted to be of much interest and importance. I hope to prove that coal can be carried on the Pennsylvania Railroad at rates that will be at the same time satisfactory to operators, and highly remunerative to the transporters, that the interests of both these classes are identified, and that the Pennsylvania Railroad Company can add millions to its receipts and hundreds of thousands to the dividends of stockholders, while at the same time increasing the revenues of the State, and conferring important benefits upon the public generally.

In the belief that the labor which has been expended upon the preparation of the following pages has not been entirely lost, and that they will be found to repay the time occupied in their perusal, they are herewith respectfully submitted for your consideration.

H. HAUPT.

Is the bituminous coal business of sufficient importance to the Pennsylvania Railroad Company to justify any efforts to retain or increase it? What is the capacity of the Pennsylvania Railroad with the present equipment of cars and engines? What would be its capacity with an increased equipment? At what cost can a coal business be accommodated, and what will it pay?

These are questions upon which great diversity of opinion exists, and in reference to which it does not appear that proper efforts have been made to determine the facts or to present them in such a simple and intelligible form as to enable those who are not familiar with intricate calculations to follow the process of reasoning and verify the results.

It is hoped that the following pages will be free from this objection. The facts and figures can generally be verified by comparison with the statistical tables accompanying the Report of the Pennsylvania Railroad Company for 1856, which the writer has had the privilege of examining by permission of the President and other officers of the Company.

The transportation business of the Pennsylvania Railroad is organized under four departments, the expenditures in which include the whole expenses of transportation; these departments are designated, Conducting Transportation, Motive Power, Maintenance of Way, and Maintenance of Cars.

The inquiries which naturally present themselves for consideration are, Can an increased tonnage in coal be accommodated with the present equipment of cars and engines, and to what extent?

How much would this increased tonnage increase the expenses in each of the departments, on the supposition that no additional equipment is required?

How much would the cost of transportation be increased if additional cars and engines should be purchased for the business?

It must be perfectly obvious that if a coal business is ever to be established on the Pennsylvania Railroad, it can only be done by transporting at such rates as will meet the competition of other localities, and it is for the Pennsylvania Railroad Company to consider what these rates are, whether they will afford the Company a profit on transportation, and if they will not afford a direct profit, whether there are any indirect or incidental advantages sufficient to justify the Company in carrying coal at rates that are not clearly remunerative. The whole subject is believed to be included in these simple interrogatories.

1. We proceed to inquire what is the capacity of the present equipment of the Pennsylvania Railroad for transporting tonnage.

The number of tons actually carried during any one month, will give the number that could be carried with the same equipment, and under similar circumstances, during every other month.

The greatest number of tons moved in any month was in April, and amounted to 45,063 tons. The number of tons that could have been moved for the same distance and in the same direction during the year, was twelve times 45,063 equal to 540,756. The number of tons actually carried was 452,992, so that, except for snow and accidental obstructions, 87,764 tons more could have been carried without an increase of equipment. The records also show that the greatest tonnage was carried during the month in which the heaviest through business was accommodated. The through tonnage east in April was 16,089 tons, the average for the year 7,392. As a much larger tonnage can be carried with the same equipment for a short than for a long distance, it is obvious that if the tonnage should be increased to the capacity of the equipment by freights carried a shorter distance, than the whole length of the road, as from the Alleghany Mountains, the tonnage capacity with a given number of cars and engines would be increased. If, then, 87,764 tons more could have been carried a distance equal to the April average, which was 323 miles, it would be proper to conclude that over 100,000 tons could be carried in the same cars if the distance were reduced to 252 miles.

It must be observed, however, that the maximum tonnage consists of freight carried both ways, whereas an increase in coal would be carried only in one direction. It becomes from this circumstance a much more difficult matter to arrive at the tonnage capacity for coal of the present equipment of cars and engines.

In this calculation the westward tonnage must be disregarded, since it is much less in amount than the eastward tonnage, and can always be accommodated in return cars.

It is also important to consider not merely the actual tonnage moved, but also the distance carried; in other words, the mileage of tonnage.

By inspecting the following table, it will be seen that the greatest number of tons was moved eastward in the month of December (28,047 tons), the average distance carried 177 miles. But the greatest mileage was made in April, 5,448,000 tons carried one mile, the average distance carried on the Pennsylvania Railroad being 219 miles.

	Tons East.	Mileage East.	Average dist.
Jan.	18,284	3,621,000	192
Feb.	15,302	2,724,000	178
Mar.	19,316	3,617,000	187
April,	24,876	5,448,000	219
May,	26,924	5,362,000	200
June,	25,061	4,648,000	186
July,	24,304	4,284,000	176
Aug.	26,113	4,521,000	173
Sept.	23,107	4,185,000	181
Oct.	27,074	4,896,000	181
Nov.	26,433	4,682,000	177
Dec.	28,047	4,961,000	177
	12)284,937	12)52,949,000	186
	23,745	average 4,412,000	

It is not to be supposed that the Pennsylvania Railroad has arrived at absolute perfection in its management; by greater dispatch in loading, unloading, and returning cars by avoiding the delays of hauling through the streets of Philadelphia, and by effecting such arrangements with the State that the load can be increased to ten tons, as allowed on other roads, and empty cars returned at any hour, it is certain that a much larger number of tons could be transported; but taking the April mileage as the measure of the absolute capacity of the car equipment, it will give an annual mileage of 65,856,000 miles. The actual mileage was 52,949,000. The difference 12,907,000. If this mileage should be employed in the transportation of coal from the Alleghany Mountain to Harrisburg, 143 miles, it would allow of an increase of 90,000 tons, and to Philadelphia, 253 miles, of 51,000 tons; as the business would be partly through and partly local, it is probable that at least 50,000 tons additional could be carried with the present equipment of cars after allowing for winter obstructions, loading and unloading, and other contingencies.

This increase is much less than the writer expected, and the results indicate a uniformity of car-movement that is commendable to the management, and must be gratifying to the directors of the Company; but when the cause of this regularity is inquired into, it appears that it must be credited entirely to coal. The coal tonnage in 1856 was 190,344 tons, or 42 per cent. of the whole tonnage of the road. Without this trade, which many consider of no value, a most discouraging and injurious falling off would have been experienced in the tonnage and receipts of the road for 1856. The coal business, through the impetus given it by the repeal of the tonnage-tax, came in most opportunely to save the credit of the Company and continue its large net earnings; exclusive of the amounts paid for tolls on other roads, the receipts from this business were nearly all profits, as will be shown hereafter. 200,000 tons can easily be added to the tonnage of last year; the question, will it pay, will be considered in its place. It is probable that without the repeal of the 3 mill-tax, not more than 50,000 tons could have been carried last year, if that much.

Having ascertained the car capacity, it is necessary to consider whether the motive-power department could carry the increased tonnage. To probe this subject thoroughly, it would be necessary to have the monthly mileage of engines on each division of the road and other data not furnished in the tabular statements. In the absence of such data, the following considerations are presented: All, or nearly all the eastward freight passes over the eastern division of the Pennsylvania Railroad. In December, the eastward movement was 28,047 tons, equivalent to 336,564 tons per annum; the actual movement for the year was 284,937. The difference which shows the increase on the supposition that the motive-power in December was worked to the limit of its capacity would be 51,627 tons, which is almost precisely the same as the increase of tonnage that the car equipment could accommodate. It must be observed, however, that there is a great difference between the motive-power required to accommodate a given eastward tonnage on the eastern and on the western division of the road; a difference of nearly three to one in favor of the former, as will be shown hereafter in considering the cost of transportation, with additional equipment.

Our next inquiry will be directed to *cost of transportation*.

In order to understand the subject under consideration, it is necessary to keep the main question prominently in view, which is this. Owing to the competition of other localities, the coal business on the Pennsylvania Railroad cannot be continued or increased if the charge for transportation exceeds certain limits. Is this limit sufficiently high to pay the Company a fair profit?

To determine this question, it is obviously necessary to ascertain how much the expenses will be increased, and it is equally obvious that the increased business is not properly chargeable with those items of expense which will be the same whether the business be accommodated or not.

In searching for data, from which to prepare an estimate, it is proper to consider only those items of expense which will be affected by the increased tonnage.

CONDUCTING TRANSPORTATION.

The items in this department are advertising, station agents, car-furniture, City Railroad tolls, clerks, conductors, depot and shop rent, fluid for lights, foreign agencies, incidentals, loss and damage, office rent and furniture, oil and tallow, repairs to buildings, stationery and printing, superintendence, teaming, firemen, brakemen and laborers, State tax, and telegraph. None of these items would be increased by an additional coal tonnage, excepting conductors, brakemen and oil for cars. There is no tax on coal, and no teaming paid by Company. Omitting the tolls on other roads, the whole freight expenses for 1856 were $702,422. The variable expenses of conductors, brakemen, and oil, &c., were $171,926.66. The expenses that would be increased by an increased tonnage constitute but 24 per cent. of the whole expenses, exclusive of tolls, in the conducting transportation department. To what extent these variable expenses would be increased remains to be determined.

We proceed to inquire, first, what was the whole expense for brakemen and conductors during the month of heaviest tonnage, and how much did it exceed the average of the year?

The expenses for brakemen and conductors in April were, for brakemen, $10,983; conductors, $3,241; total, $14,224. The average for the year was, for brakemen, $10,233; conductors, $2,930; total, $13,163.

During the month of greatest tonnage the expenses for conductors and brakemen exceeded the average $1,061. The tonnage exceeded the average in the same month 7,314 tons, or 14 cents per ton; the average mileage was 328 miles, and the increased expense in these items consequent upon the increased tonnage was $\frac{42}{100}$ of a mill per ton per mile.

The whole expenses for oil and tallow for freight cars was $13,971; the whole mileage 119,836; 501 tons carried one mile, the cost per ton per mile $\frac{11}{100}$ of one mill, which exceeds the cost of the same item in coal transportation on the Reading Railroad $\frac{3}{100}$ of a mill.

It appears, therefore, that the whole increased expenses in the department of conducting transportation would be covered by $\frac{53}{100}$ of a mill per ton per mile.

The whole expense of all the variable items in the department of conducting transportation having been $171,926 66, and the whole mileage 119,836,501 tons carried one mile, the whole expense of these items on the whole business of the year was $1\frac{43}{100}$ of a mill per ton per mile. This result shows that an increased tonnage carried without an increased equipment, has been accommodated at 37 per cent. of the average expense of the variable items.

MAINTENANCE OF WAY.

Were the maintenance of way expenses greatest when the heaviest tonnage was carried on the road, and how much did they exceed the average?

A large portion of the maintenance of way expenses are independent of the tonnage carried, such as ties, telegraph, repairs of hand-cars, tool and workmen's houses, bridges, removal of snow, taxes on real estate, watchmen, repairs of tools. These items are the same, or nearly the same, whatever the tonnage may be. The only items of expense in this department which can be supposed to be affected by an increase in the tons carried, are labor and material for repairs of track and iron rails.

The month of April, in which the heaviest business was done, is also the month when, from the effects of the winter frost, it might be expected that, without any increase in the business of the road, the repair-expenses would be at the highest point. During the month of January, but little work can be done on the track, and the expenses were $2,548; in April, they were $14,388. The average for the whole year was $11,704. Leaving out the winter months of December, January, February, and also March, during which but little work is done on track-repairs, the average for the balance of the year was $14,631, which exceeds the expenses in April; and the expenditures in May, June, August, September and October, were each greater than in April, although the tonnage and mileage were less. It thus appears that the monthly expenditures for labor and materials for track are not affected perceptibly by a moderate increase of tonnage.

The only item seriously affected by an increase of tonnage is the wear of rail. In examining the monthly statement, it appears, as might have been expected, that the monthly expenditures for rails bear no proportion to the tonnage. They vary from $8 to $15,738 in one month. The Pennsylvania Railroad is as yet too new to furnish any reliable data on this subject. The only road which exhibits, for a period of years, the effect of tonnage on rails is the Reading Railroad, where this important subject has received the attention it deserves. The average cost of renewing rails for the last nine years, from 1848 to 1856, inclusive, has been $2\frac{82}{100}$ cents per ton, over the whole road, including the wear from return cars and engines. The distance being 95 miles, the cost per ton per mile was $\frac{31}{100}$ of a mill. The greatest expenditure for renewals in any one year, when a large amount of new track was laid, was $\frac{8}{10}$ of a mill per ton per mile. One-third of a mill per ton per mile, from the experience of the Reading Railroad, would appear to be a fair average allowance for wear of rails, and this may be reduced, by improving the quality of the iron and the mode of securing the joints.

At this rate, the wear of rails on the Pennsylvania Railroad, from the total tonnage of 84,777,214 tons carried one mile, would be $28,259. The actual expenditure for this item, in 1856, was only $16,031. The whole expenses of maintenance of way were $324,737. The only item sensibly affected by an increased tonnage being wear of rails, the percentage of the variable expenses in this department would be less than 9 per cent. In other words, only 9 per cent. of the maintenance of way expenses, on the Pennsylvania road, would appear to be increased by increasing the tonnage, the other items being very nearly constant.

The next inquiry is, do the expenses of repairs of freight cars increase in proportion to the tonnage, and what are these expenses per ton per mile?

MAINTENANCE OF CARS.

As the repairs of cars are not made or the expenses of repairs charged until after the accidents, or wear which rendered the expense necessary, it is impossible to trace the connection between the tonnage carried and the wear which it occasions. For example, the expenditure for car repairs on the Pennsylvania Railroad was greatest in the month in which the smallest tonnage was carried, and was below the average in the month in which the greatest tonnage was carried.

It is the opinion of many of the most experienced railroad managers, that cars deteriorate nearly as much when standing exposed to the weather as when in use. The woodwork will certainly rot as fast, and the iron will rust more rapidly where the rubbing parts are not lubricated.

It is also obvious that if cars are not fully loaded, the addition of a weight sufficient to make up the deficiency will not add appreciably to the expense of car repairs. For these reasons, it is proper to conclude that an increased tonnage carried in a given number of cars, cannot increase proportionably the expenses of car repairs, but to avoid the possibility of making too low an estimate of these expenses, all these considerations will be left out of view, and the increased tonnage will be charged with the average of the whole business.

The whole expenditure for repairs of freight cars on the Pennsylvania Railroad for 1856, has been $134,637. The number of tons carried one mile, 119,836,501, the expenses per ton per mile would be $1\frac{1}{10}$ mill.

On the Reading Railroad, the cost of repairs of freight cars per ton per mile has been $1\frac{1}{5}$ mill, and for coal cars $\frac{96}{100}$ of a mill per ton of 2240 lbs. per mile.

It may very safely be assumed, therefore, that for an additional business on the Pennsylvania Railroad, even with an increase of equipment, $1\frac{1}{10}$ mill per ton per mile would be an ample allowance for the department of maintenance of cars.

MOTIVE POWER.

In what proportion would the motive power expense be increased, by an increase in the tonnage carried?

The items which constitute the motive power expenses, are depot and shop rent, fluid for lights, incidentals, oil and tallow, repairs to buildings, stationery and printing, superintendence, laborers, watch and switchmen, fuel, cotton waste, engineers and firemen, expenses of water stations, fuel for stationary engines, repairs of locomotives, repairs of tools and machinery.

Although, at first view, it might be supposed that the motive power expenses would increase nearly in the same proportion as the tonnage, yet it is obvious, from an inspection of the list, that many of these items would not be affected by an increased tonnage in coal. Fuel, oil and tallow, cotton waste, engineers and firemen, and repairs of locomotives, are the only items that would be sensibly affected by any increase of tonnage, that did not require new buildings and new shop machinery to accommodate it, and even then the increase per ton per mile would be very trifling.

From an examination of the records, the following facts are elicited. The whole expenditures in the motive power department during the month of April, when the heaviest tonnage was carried, were less than the average of the year, and less than in the month in which the least tonnage was carried.

The whole motive power expenses during the year, in the freight department, amounted to $354,866. The mileage on the Pennsylvania and Lancaster and Harrisburg Railroad, 94,107,576 tons carried one mile. The total expenditure per ton per mile $3\frac{7}{10}$ mills.

Taking the items which are affected by an increased tonnage, it appears that the expenditure for coal in April was $4,574, the average of the year $4,200. The number of tons carried during this month exceeded the average 7,314 tons, the average distance carried was 248 miles, the cost of coal for the excess over the average was therefore $\frac{21}{100}$ of a mill per ton per mile.

The expenditure for coal on the whole business of the year was $50,401. The mileage of freight hauled by locomotives of Pennsylvania Railroad Company, 94,107,576. The cost per ton per mile $\frac{53}{100}$ of a mill.

It appears from these results that an increased tonnage, without increase of equipment, has added to the consumption of fuel per ton per mile carried, less than half the average consumption on the whole business.

The pay of engineers and firemen, in the month of greatest tonnage, was $6,707; in the month of least tonnage it was $7,167; the average was for the whole year $6,314. It does not appear, therefore, from the statistics of last year's business, that the monthly fluctuations in tonnage have affected in any degree the expenditures for wages of engineers and firemen.

The total disbursement for engineers and firemen, for the year in the freight department, was $75,766. The mileage being 94,107,576 tons one mile, the expense per ton per mile has been $\frac{8}{10}$ of a mill.

The next item is repairs of locomotive engines, and here, as might be expected, there is an entire absence of data by which to determine how much the expenditures in this department would be increased, by a moderate increase of tonnage carried within the limits of the power, and without an increase in the number of the engines. That the wear of an engine with a full load, would be something more than with a light one is obvious, but no one can suppose that the expense of repairs can increase in the same proportion as the tonnage. The monthly record throws no light on this subject. During the month of heaviest tonnage, the repairs of freight engines cost $15,428, or less than the average of the year, which was $15,616 and much less than in the month of lightest tonnage, which was $17,834. The whole cost of repairs of freight engines for the whole year, was $187,382. The number of tons carried one mile on the Pennsylvania Railroad and Harrisburg Railroad 94,107,567, the cost per ton per mile $1\frac{9}{10}$ mills, which, for the reasons given, is obviously much more than would be incurred in the transportation of an increased tonnage within the capacity of the equipment.

These results are based upon an average of the whole business of the Pennsylvania Railroad on both the Eastern and Western Divisions, and in the motive power department are greatly in excess when applied to a local business transported over a line of descending grades. The motive power expenses on each division, when estimated per ton per mile, differ very greatly. On the eastern division the expenses of engines for 100 miles run, have been for 1856:—

Repairs, $6 75; fuel, $8 09; stores, $1 53. Total, $16 37. On western division:—

Repairs, $11 36; fuel, $7 82; stores, $1 95. Total, $21 13.

As the object of this examination is to determine the cost of carrying coal eastward, the capacity of an engine on each division must be considered. On the western division a first class engine could carry an average load besides cars of 125 gross tons, the ruling gradient being $52\frac{8}{10}$ feet to the mile. On the eastern division, where the grades are generally level or descending, the same engine can carry 325 tons. If then the cost per 100 miles run on the eastern division be $16 37, and the number of tons hauled 325, the cost per ton per mile will be $\frac{5}{10}$ of a mill, and doubling this result, which is much more than a proper allowance, for the cars returning empty, will give as the whole cost of fuel, repairs, and stores on eastern division, one mill per ton per mile.

To this must be added wages of engineers and firemen, which are thus determined.

The whole number of miles run with freight trains has been 1,377,284. The whole expenses of engineers and firemen $75,776. The average expenses per mile run over the whole road $5\frac{1}{2}$ cents, which divided by 325, the load on the eastern division, will give per ton per mile $\frac{17}{100}$ of a mill; double this result for the return engines with empty cars will make $\frac{34}{100}$ of a mill per ton per mile.

On the western division the expense of fuel, stores, and repairs, being $21 13 per 100 miles, and the number of tons hauled to each train 125, will give per ton per mile $1\frac{7}{10}$ mills, which being doubled to include return trains, makes $2\frac{3}{10}$ mills per ton per mile.

The item of engineers and firemen both ways being 11 cts. per mile, and the load of a train on western division 125 tons, the cost of this item will be $\frac{9}{10}$ of a mill per ton per mile.

GENERAL SUMMARY OF RESULTS.

1. With the present equipment of cars and engines, the tonnage of the Pennsylvania Railroad between the summits of the Alleghany Mountains and Philadelphia, might be increased from fifty to sixty thousand tons.

2. Within this limit of 50,000 tons an increased business could be carried at the following cost:—

Conducting Transportation.

Brakemen and conductors per ton per mile	0.42 mills.
Oil and tallow for cars " "	0.11 "
Total	0.53

Maintenance of Way.

Wear and renewals of rails per ton per mile, 0.33 mills.

Maintenance of Cars.

Repairs of cars, 1.10 mills.

Motive Power.

Fuel per ton per mile	0.21 mills.
Engineers and firemen	0.12 "
Repairs	0.20 "
Stores	0.05 "
	0.58

The above items in the motive power department, it will be observed, are intended to apply only to an increased tonnage hauled with the present equipment of engines. They are set down at higher figures than the monthly variations of expenses would indicate, as some of their expenses have been least where the tonnage was greatest.

The aggregate of all the expenses is only $2\frac{54}{100}$ mills per ton per mile.

The conclusion to which we are led is, that 50,000 tons of

coal, in addition to the present business, could be carried over the Pennsylvania Railroad from the summit of the Alleghany Mountains to Harrisburg, at an actual cost to the Company of less than 3 mills per ton per mile on that part of the road owned and operated by the Pennsylvania Railroad Company.

This conclusion will be regarded as an absurdity by those who are determined to estimate the cost of transportation in no other way than by taking the expenses of the whole business of a road on a small tonnage; but a moment's reflection must satisfy any person of ordinary intelligence that there are very many items of permanent expense in a railroad that are not at all affected by an increase of business, and it should require very little argument to prove that if trains are run with the cars half loaded the expense will be very nearly the same as if the cars were filled. To the extent of the capacity of the equipment of cars and engines, there is nothing remarkable in this result.

Beyond the limit of 50,000 tons the circumstances are different; for the accommodation of this increased business additional cars and engines must be provided. It is obvious that there are still many items of expense which this additional business will not affect, and that it should cost much less than the ordinary business of the road which is chargeable with the constant as well as the variable expenses. To determine the cost of this business, the following items are presented:—

Conducting Transportation Department.

Aggregate of all the variable expenses per ton per mile . . . . 1.43 mills.
Maintenance of way . . . . . 0.33 "
Maintenance of cars . . . . . 1.10 "

Motive Power on Eastern Division.

Fuel, repairs and stores . . . . 1.00 mills.
Engineers and firemen . . . . 0.34 "
                                                        1.34

Motive Power on Western Division.

Fuel, repairs and stores	2.30
Engineers and firemen	0.90
	3.20

Total cost of carrying coal on the eastern division with an additional equipment of cars and engines, and exclusive of interest on capital, per ton per mile 4.20.

On the western division, the total cost per ton per mile would be 6.06 mills.

These figures are believed to be reliable, but it must not be forgotten that they bear no necessary relation to the average expenses of the regular business. In the estimate that has been made, all these constant expenses which are not increased by an increase of tonnage have been left out of view entirely, and this is obviously proper, for the reason that the question to be determined, is simply at what increased cost can a proposed increased business be accommodated, in order to determine whether the business will pay remunerative rates within the limits that can be afforded by the shippers or operators.

These results present nothing impossible or unreasonable. The cost of carrying coal on the Reading Railroad, where it is charged with its proportion of the constant as well as the variable expenses, is only 6 mills per ton per mile, and the gross ton at that. The cost on the Eastern Division of the Pennsylvania Railroad, with an equal amount of tonnage, should not be made greater.

It is improper to compare rates charged on different roads without taking into consideration the difference of circumstances. The present charges on the Baltimore and Ohio Railroad, and some other roads, are often cited as a reason why the rates on the Pennsylvania Railroad should not be reduced. It is said we are now carrying at lower rates per mile than other roads. To those who look merely at the surface of things, this kind of reasoning may appear plausible; but it must be remembered that the Baltimore and Ohio Railroad is a road of high grades and sharp curves, with only 44 miles of second track east of Cumberland, and a coal tonnage, considerably exceeding half a million of tons, a road which has consequently reached its capacity for transportation, and in the desire of the Company to accommodate its large and increasing through business, the charges on coal have been necessarily increased. For many years the Baltimore and Ohio Company carried coal at less than one cent per ton of 2,240 lbs. per mile, and it was declared to be remunerative even on that road by those who were most intimately connected with its management. From the results of a careful and detailed investigation of the effects of grades, by J. Dutton Steele, Esq., Civil Engineer, published in the Report of the Reading Railroad Company for last year, it is very clear that the Baltimore and Ohio Railroad has nearly reached the limit of its capacity on the single track portions, and the Company does not desire, and could not accommodate an increased coal business, if they did desire it, until the double track is laid between Cumberland and Baltimore.

The Pennsylvania Railroad is in a different position. The capacity of the road between the Alleghany Mountain and Columbia, with its long turnouts and portions of double track, its descending grades, and the use of the telegraph, may be considered from the same authority as nearly equal to a million of tons. Its capacity is sufficient to transport as much as the Columbia Railroad can, with its other business, remove. The high tolls on the State road are a serious drawback, but as a new outlet will soon be afforded by the Lebanon Valley Railroad, it is probable that arrangements can be made to send cheaply and expeditiously to Philadelphia all the coal that reaches Harrisburg; no doubt the Reading Railroad Company would carry at the average rate per mile charged for the whole distance.

On the Philadelphia and Columbia Railroad, the Pennsylvania Railroad Company provide cars and pay conducting transportation expenses, the State provides road and motive power. The increase in the motive power expenses on this road from an increased tonnage should not be as great as on the Western Division of the Pennsylvania Railroad, where the Alleghany Mountain is to be overcome with 53 feet grades, and should, therefore, be less than $3\frac{2}{10}$ mills per ton per mile, which includes return cars empty. The wear of rails being $\frac{33}{100}$ of a mill, the whole cost of an increased business on the Columbia Railroad should be but $3\frac{53}{100}$ mills per ton per mile, or about 29 cents over the whole road; even if maintenance of way expenses be set down at $\frac{2}{3}$ of a mill, the cost to the State would be but 32 cents, including return cars empty. When charged with the constant expenses, in addition, the cost of the ordinary business is reported at about $77\frac{1}{10}$ cents.

It may be interesting, in connection with the transportation over the State road, to inquire into the relative proportion of the maintenance of way and motive power, and the maintenance of cars, and conducting transportation expenses on the Pennsylvania Railroad for 1856.

The whole motive power expenses of the freight department were $354,866. The mileage of freight over Pennsylvania Railroad and Harrisburg Railroad, 94,107,576 tons one mile; the whole cost per ton per mile, $3\frac{7}{10}$ mills.

The whole maintenance of way expenses on Pennsylvania Railroad, were $194,842; the mileage 84,777,214 tons, one mile, the cost per ton per mile, $2\frac{3}{10}$ mills. The sum of these two items is 6 mills.

The conducting transportation expenses of the freight department, exclusive of tolls on other roads, were $702,422.04. The mileage, 119,836,501. The cost per ton per mile, $5\frac{8}{10}$ mills.

The maintenance of car expenses were $139,004; the mileage, 119,836,501 tons one mile, the cost per ton per mile less than $1\frac{2}{10}$ mills.

The sum of the last two items is 7 mills.

The average of the whole freight business of the Pennsylvania Railroad, obtained by adding these items, is $1\frac{3}{10}$ cents per ton per mile.

The car and transportation expenses have exceeded the sum of the road, and motive power, by 16 per cent., having been in the proportion of 7 to 6.

This proportion, however, does not indicate the equitable division of receipts over the Columbia Railroad, as the amount of capital in the different departments of the service forms an essential element in the calculation.

As this subject is one of great interest to the Pennsylvania Railroad Company, the writer will endeavor to ascertain the proportions in which the freight receipts should be divided over the Columbia Railroad in the ordinary business. The report of the Canal Commissioners for 1856 not being yet received, the calculations will be based on the report of 1855:—

The receipts of the Columbia Railroad from freight were 67 per cent. of the whole receipts.

The mileage of freight, 382,662 tons 81 miles $\doteq$ 30,795,946 tons one mile.

The whole cost of motive power was	$324,008
The cost, per trip	$30
The average number of freight cars hauled per trip	24
Proportion of motive power for freight	$234,680
Whole cost of motive power per ton per mile	$7\frac{6}{10}$ mills.

It must be observed that the average number of cars hauled was only 24, whereas a first class engine could haul nearly double that number with scarcely any increased expense, when distributed per ton per mile over the whole road.

The maintenance of way expenses were $104,300
Proportion due to freight 69,881
Per ton per mile for freight in mills . . $2\tfrac{1}{10}$
Taking the cost of the road, without equipment, at 4,000,000, the interest would be $240,000, and the proportion applicable to freight $160,000
26 freight engines, with depot room and tools, $260,000
    Interest 15,600
                                              $175,600
Interest on capital per ton per mile . . $5\tfrac{8}{10}$ mills.

It must be observed here also that the tonnage of the Columbia Railroad was only 382,662 tons, whereas the capacity of a double track road with a ruling gradient of 45 ft. exceeds one million of tons in addition to passenger business. All additional tonnage would greatly reduce this allowance per ton per mile for capital.

Adding together the items for the State road, the aggregate is—

    Motive power . . . . . 7.6 mills.
    Maintenance of way . . . 2.2 "
    Capital . . . . . . 5.8 "

    Total . . . . . . 15.6 "

If the tonnage should be doubled, these figures would be reduced to about 10 mills.

It has been already shown that the conducting transportation expenses of the Pennsylvania Railroad Company amount to $5\tfrac{8}{10}$ mills per ton per mile, and for maintenance of cars $1\tfrac{2}{10}$ mills, total, 7 mills.

To this must be added, as in the case of the State road, the interest on the capital invested in these departments, an approximate estimate of which is as follows:—

Real Estate required by Transportation Business.

On the road	$400,000
Pittsburg	500,000
Philadelphia	700,000
Total,	1,060,000
Car repair shops and machinery	300,000
Station and warehouses	800,000
Freight cars	800,000
Telegraph	29,000
To which must be added subscriptions to unproductive roads, designed to bring business on the whole line,	1,200,000
	$4,729,000
Annual interest	283,740
Chargeable to freight	210,000

Amount per ton per mile, 119,836,501 miles, $1\frac{8}{10}$ mills.
Adding these items, which represent expenses of Pennsylvania Railroad Company over State road,

Conducting transportation	5.8	mills.
Maintenance of cars	1.2	"
Capital account	1.8	"
	8.8	"
Adding for road and motive power	15.6	
Total	24.4	"

If the gross receipts should be divided between the Pennsylvania Railroad Company and the State in the proportion of these figures, the State should receive 64 per cent., and the Pennsylvania Railroad Company 36 per cent.

If the trains had carried the full complement of cars for which an increased tonnage in coal would have offered the facilities, the motive power and road expenses per ton would have been reduced, and the proportion of the Pennsylvania Railroad Company increased. It is probable that the allowance of 40 per cent. which the Pennsylvania Railroad Company have claimed as their due on the State road, is as near the equitable proportion as it can be ascertained.

These remarks and the conclusions above stated apply to the ordinary freight tonnage over the Columbia Railroad, and not to an increase in such an article as coal.

It is unfortunate that when Pennsylvania is so greatly indebted for her eminent prosperity to the coal and iron interests, efforts should be made by parties connected with the management of the State works to discourage the increase of the bituminous coal trade by representations which any intelligent citizen of the State with a memory reflective can see to be entirely erroneous. Is it not, for example, a simple absurdity to assert that if an addition be made to the business of the Columbia Railroad of 100,000 tons of coal, the increase will cost as much as the former average of the whole business of the road? Yet such assertions are continually made, and upon such calculations of the expenses the Canal Commissioners have repeatedly based their refusals to make reductions which the protection and increase of the trade of the State imperatively demanded.

The proper manner of making a calculation of the cost of carrying coal on the Columbia Railroad in addition to the present business, and the required data are furnished in the preceding pages, based upon the expenses on the Pennsylvania Railroad, but it may be satisfactory to make the calculation with the figures given in the reports of the Columbia Railroad as far as they are applicable.

In the maintenance of way department it has been ascertained from the reports of the Reading Railroad Company that $\frac{33}{100}$ of a mill per ton per mile is a proper allowance for wear of rail.

The other items of track expenses are not perceptibly increased by an increased tonnage, but call them equal to wear of rails, which will make the whole road expenses $\frac{66}{100}$ mill per ton per mile.

A locomotive of the first class costs on the Columbia Railroad, $30 per trip, double this sum for returning with empty cars, which is much more than a proper allowance; allow also full trains of 45 cars, and 180 tons east and empty west. Also 100 trips per annum to each engine each way, and one-third of the time in shop. This would give a full allowance for the motive power expenses. The results are 18,000 tons to each engine, at a cost of 30 cents per ton.

Add interest on cost of engine, 3 cents per ton, 33 cents.
Add again for road and wear of rails, 81 mills, $5\frac{1}{2}$ "

Total cost to the State of carrying coal, $38\frac{1}{2}$ "

This is on the supposition that new engines are provided for all the increased business; but as the engines hauled only 24 cars as an average train, it is probable that 100,000 tons more could be hauled without increasing the number of engines, and at a cost of less than thirty cents per ton, instead of 77, as has been represented. These calculations are too easily made to excuse any one in permitting himself to be misled by the statements of expenses so frequently given in official reports.

The proper division of receipts between the Pennsylvania Railroad Company and the State, for coal transported over the Columbia Railroad, does not appear to rest upon precisely the same principles which govern the division of the ordinary freight receipts. In this calculation, the fixed capital should not enter as an element, and it would appear proper that the receipts on this increased business should be divided in proportion to the variable expenses of each party, including the variable capital which it requires. If this basis be admitted, the following figures will indicate the proportions:—

For State Road.

Motive power	$7\frac{6}{10}$ mills.
Maintenance of way	$2\frac{2}{10}$ "
	$9\frac{8}{10}$ "

For Pennsylvania Railroad.

Conducting transportation	$5\frac{8}{10}$ mills.
Maintenance of cars	$1\frac{2}{10}$ "
	7 "

Divided in proportion of $9\frac{8}{10}$ and 7, will give the State 58 per cent. and the Pennsylvania Railroad Company 42 per cent.

It would appear, therefore, that a division, in the proportion of 60 per cent. to the State, and 40 per cent. to the Pennsylvania Railroad Company, would give the Columbia Railroad a very full proportion on all the freight business.

The coal operators of Pennsylvania ask only to be allowed to deposit their coal at tide water as cheaply as coal from other localities is deposited by other improvements. This is a simple and reasonable request. Without this no sales can be made, no coal can be mined, and the capital invested will be lost. The lowest rates ever asked by operators will leave large profits for the Pennsylvania Railroad Company, and for the State, if an equitable division be made between these parties.

The cost of carrying coal from the Alleghany Mountains to Philadelphia can be readily determined from the data already furnished. In making the calculations, it will be assumed that additional cars and engines are required for the whole business.

From the summit of the Alleghany Mountains to Harrisburg, 142 miles, at $4\frac{2}{10}$ mills	59.64
Harrisburg to Columbia, 29 miles, motive power, 1.34; cars, 1.10; conducting transportation, 1.43; } 3.87	11.22
Conducting transportation and cars on State road, $2\frac{53}{100}$ mills, 80 miles	20.30
Total, exclusive of tolls on other roads	91.16
The tolls on the State Road are	85.00
Harrisburg and Lancaster Railroad, say	20.00
	196.16

This amount of $1.96, includes the whole actual cost per ton, of transporting coal from the Alleghany Mountains to Philadelphia, excepting the interest on new cars and engines.

An eight wheel coal car, costing $500, and making an average trip once a week, with 8 tons, would give $7\frac{1}{2}$ cents per ton, as the interest on capital in new cars. An engine, costing $10,000, carrying 300 tons, 4 days to Columbia and back, idle one-third of the time, will carry 15,000 tons per annum, and will give 4 cts. per ton as the allowance for interest on capital. Adding $11\frac{1}{2}$ cts. to $1.96\frac{1}{8}$, gives $2.07\frac{5}{8}$ as the whole cost of carrying coal from the summit of the Alleghany Mountains to Philadelphia, with new cars and engines, and including tolls on other roads, and interest on new capital, a sum which, when the roads between Harrisburg and Philadelphia consent to an equitable division of receipts, can be much reduced.

If the road was extended to the Delaware front, and facilities provided for immediately unloading and returning cars, the business that could be accommodated with a given equipment would be much increased. If the Columbia Railroad were worked in connection with the Pennsylvania Railroad as one road, and no detention allowed at Columbia, it would require but 24 hours, at 10 miles per hour, between Altoona and Philadelphia; in this case, the round trip could readily be made in 4 days instead of six, which would add 50 per cent. to the car capacity; if, in addition to this, the cars should be allowed to carry 10 tons instead of 8, as on other roads, there would be a further increase of 25 per cent. to the capacity of a given number.

The number of cars required to carry 100,000 tons in one year running 200 days, loading 8 tons, and the round trip occupying one week, would be 375 cars, costing at $500 per 8 wheel car, $187,500. If the round trip can be made in 4 days, and the load 10 tons, running 200 days, 200 cars will accommodate the business, and the cost of cars be reduced to $100,000. At this rate the capital invested in new cars would be repaid to the Pennsylvania Railroad Company by the profits of the first year.

One more feature of this business requires a moment's consideration. An examination of the tables published by the Reading Railroad Company shows a steady and comparatively uniform increase in the passengers and tons of merchandise corresponding with the increase of coal, the passengers being about 5, and the tons of merchandise 10 per cent. of the tons of coal. If this law be applied to the Pennsylvania Railroad, it would give for each 100,000 tons of coal, 5,000 passengers, and 10,000 tons merchandise carried an equal distance; estimating passengers at $2\frac{1}{2}$ cts. per mile, and merchandise at an average of 40 cents per 100 lbs., would give an increase of receipts of $111,000 for each 100,000 tons, paying nearly half the cost of coal transportation, and almost sufficient to pay for the coal cars in a single year. The expense of accommodating this increased business except for tolls on the other roads would be almost nothing. These results would indicate that an expenditure for coal cars would return nearly 200 per cent. per annum in direct and indirect profits on the capital thus invested.

Freight carried westward in the return cars costs the Company so nearly nothing that it is difficult to estimate it. The only source of expense would be in the delay of loading and unloading; allowing one day for loading and one for unloading, counting also 200 running days to the year, and cost of car $600. The value of the time lost would be $4\frac{1}{2}$ cents per ton for the whole distance carried. This applies to lumber carried westward between Tyrone and Pittsburg, the receipts from which are nearly all profit.

Eastward the cost of lumber would be about the same as coal, but the better qualities of pine and oak could afford to pay a higher rate, and the charges on hemlock should be reduced if the business is desired.

The facts of the case have now been presented simply and truthfully as they are believed to exist. There has been no desire to underrate expenses or exaggerate profits. It is probable that from the peculiar position which the writer has occupied in reference to the Pennsylvania Railroad Company, and to operators, he has thought more and figured more than any other person connected with the road upon the cost of transportation and the ways and means of increasing the business and profits of the Company. When he first recommended to others investments of capital in coal operations, he had no expectation of being personally identified with them. Such connection was the result of a subsequent determination, based upon the belief that his own interests and the interest of the road could be better advanced by vacating the office of superintendent, and directing himself to the organization of such companies. The tonnage of the Eastern Division of the Pennsylvania Railroad eastward has been only about one-fourth of a million of tons. Six or seven hundred thousand tons more could be carried, and yet be within the limits of the capacity of the road, if additional cars and engines be provided for it. The coal business could be increased in two or three years to 500,000 tons, paying not less than $600,000 direct profit, and perhaps $400,000 more in the increase of passengers and other freight consequent upon it. It is for the Pennsylvania Railroad Company to consider, with the facts before them, whether this business is worthy of encouragement. To accommodate it, an extension to the Delaware point is obviously indispensable. Whether the extension be made to League Island, Greenwich or Richmond is of comparatively little consequence, so that ample room is afforded for a business of such magnitude with steam communication. It is also obvious that as the capacity of the present equipments of cars and engines has been nearly reached, there can be but little increase without an additional supply, which, to have an influence on the business of 1857, cannot be procured too soon.

It has been said, "admitting that the cost of an increased coal business on the Pennsylvania Railroad is not more than has been represented, the Pennsylvania Railroad Company seek to obtain from each branch of business as high rates as it will bear." Influenced by such principles, the officers of the Company make their own estimates of the cost of mining, fix their own market value upon the coal, assign what they consider sufficient profit to keep the business alive, and charge all that remains for freight. The contingencies of business are not included in the estimate; while it is impossible to find any responsible and experienced party willing to sell coal, and guarantee sales for less than 25 cts. per ton, the Pennsylvania Railroad Company do not calculate this service as worth anything, but have sometimes given operators to understand that 25 cents should be sufficient to cover cost of coal in mines, profits on operation, guarantees of sales, losses in transportation and reshipments, incidentals and all the contingencies of business; even in these calculations the cost of mining is estimated too low, and the market value too high, and the result is that operators seeing before them nothing but a certainty of loss if they make contracts without a fixed and satisfactory rate of charge for transportation, are deterred from doing anything, the season for contracts passes unimproved, and other States and foreign collieries supply the market.

A rule for fixing rates has been suggested to the Pennsylvania Railroad Company, as that which is at the same time most simple and equitable. At whatever cost for transportation coal can be delivered from the Cumberland mines in Philadelphia, New York, Boston, or other points, let such rates be established on the Pennsylvania Railroad, as will permit Pennsylvania coal to be delivered in the same places at the same cost. This would be a safe rule, for the Pennsylvania Railroad Company can transport cheaper than any of its competitors, and a proper representation of the facts will no doubt lead to equitable charges on the State road. As to the Harrisburg Road there can be no difficulty; the cost of coal transportation to this company being less than 2 cents per ton, it is not to be supposed that they would resist any reduction necessary to secure the business.

It is respectfully suggested, also, that the principle of exacting from every business that passes over the Pennsylvania Railroad the highest rates that it will bear, is not that which influenced the founders and originators of this great improvement. The increase of trade, the extension of Philadelphia, the development of the agricultural and mineral resources of the State, the opening of mines and collieries, the establishment of factories, the increase in the population and in the value of property along the road, were all prominent objects when the road was started, and were assigned as reasons for its construction more powerful than the expectation of large dividends; but so far as coal is concerned, all these beneficial results flow from the same source, a large business and large dividends go together. The profits on coal are so small that a slight reduction just sufficient to give a profit to the operator may quadruple the shipments, and increase tenfold the net profits of the transporter.

Printed in Poland
by Amazon Fulfillment
Poland Sp. z o.o., Wrocław